THE SCIENCE OF NATURAL DISASTERS

THE SCIENCE OF
BLIZZARDS

Joanne Mattern

Cavendish Square

Published in 2020 by Cavendish Square Publishing, LLC
243 5th Avenue, Suite 136, New York, NY 10016

Copyright © 2020 by Cavendish Square Publishing, LLC

First Edition

Website: cavendishsq.com

This publication represents the opinions and views of the author based on his or her personal experience, knowledge, and research. The information in this book serves as a general guide only. The author and publisher have used their best efforts in preparing this book and disclaim liability rising directly or indirectly from the use and application of this book.

All websites were available and accurate when this book was sent to press.

Library of Congress Cataloging-in-Publication Data

Names: Mattern, Joanne, 1963- author.
Title: The science of blizzards / Joanne Mattern.
Description: First edition. | New York : Cavendish Square, 2020. |
Series: The science of natural disasters | Audience: Grades 2 to 5. |
Includes bibliographical references and index.
Identifiers: LCCN 2018055116 (print) | LCCN 2018056795 (ebook) |
ISBN 9781502646415 (ebook) | ISBN 9781502646408 (library bound) |
ISBN 9781502646385 (pbk.) | ISBN 9781502646392 (6 pack)
Subjects: LCSH: Blizzards--Juvenile literature. | Winter storms--Juvenile literature.
Classification: LCC QC926.37 (ebook) | LCC QC926.37 .M3845 2020 (print) | DDC 551.55/5--dc23
LC record available at https://lccn.loc.gov/2018055116

Editorial Director: David McNamara
Editor: Kristen Susienka
Copy Editor: Nathan Heidelberger
Associate Art Director: Alan Sliwinski
Designer: Ginny Kemmerer
Production Coordinator: Karol Szymczuk
Photo Research: J8 Media

The photographs in this book are used by permission and through the courtesy of: Cover Ipedan/Shutterstock.com; p. 4 Zorah Ras/Shutterstock.com; p. 5 (background, and used throughout the book) Lightkite/Shutterstock.com, Sarkelin/Sjutterstock.com; p. 6 hxdbzxy/Shutterstock.com; p. 7 WoodysPhotos/Shutterstock.com; p. 8 lev radin/Shutterstock.com; p. 9 robypangy/Shutterstock.com; p. 11 geno4ka01/Shutterstock.com; p. 12 Handout/Getty Images; pp. 15, 22, 25 Boston Globe/Getty Images; p. 17 Sylwia Duda/Moment/Getty Images; p.18 Evannovostro/Shutterstock.com; p. 19 John Wollwerth/Shutterstock.com; p. 27 ©AP Photo; p. 28 Miriam Doerr Martin Frommherz/Shutterstock.com.

Printed in the United States of America

CONTENTS

A blizzard can bury cars and sidewalks and make travel almost impossible.

CHAPTER 1

WHAT IS A BLIZZARD LIKE?

In cold weather, snow can fall. Snow can fall fast or slow. Sometimes the wind moves the snow around too. Sometimes a blizzard can form. A blizzard is a special kind of weather event. It happens when wind and snow combine.

Is It a Snowstorm or a Blizzard?

A blizzard and a snowstorm are different. A snowstorm needs three things to form. It needs cold air and

Dark clouds tell people a storm is coming.

moisture. It also needs **lift**. Lift happens when cold air and warm air meet. The warm air rises above the cold air. This changes the moisture in the air to snow. Sometimes just a little snow will fall. Sometimes a lot will fall. It all depends on how much moisture is in the storm clouds.

To be called a blizzard, a storm must have four things. First, there must be heavy snow. Next, the wind must blow at 35 miles per hour (56 kilometers per hour) or more. Third, **visibility** must be less than 0.25 miles (0.4 kilometers). Finally, the storm has to last for three hours or longer.

Blizzard Country

Blizzards happen in most of the northern parts of the world. Examples are Russia, Canada, and parts of northern Europe. Parts of the United States have blizzards too. In the United States, most blizzards strike the northeastern and north-central parts of the country.

Strong winds can blow snow sideways and create blinding conditions during a blizzard.

Wherever blizzards happen, they can mean a lot of danger. A blizzard's extreme conditions can even cause death.

A 2018 blizzard in New York City buried city streets.

Dangerous Conditions

Blizzards are dangerous in many ways. Snow can make roads slippery. Blowing snow makes it hard to see. There

Blizzards can make roofs collapse because the snow is so heavy.
This person is shoveling snow off of a roof that has collapsed.

can be car accidents. People can be trapped if cars get stuck in the snow. Heavy snow can make roofs collapse.

DID YOU KNOW?

The worst US blizzard struck New York City and the East Coast in 1888. The storm lasted for two to three days. It dropped up to 50 inches (130 centimeters) of snow. Winds blew at more than 45 miles per hour (72 kmh).

WHERE DO BLIZZARDS HAPPEN?

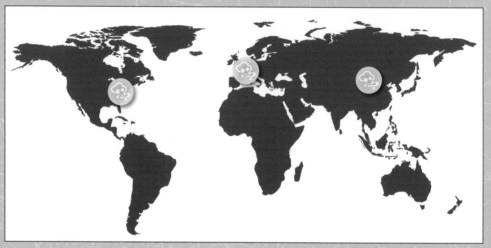

The icons on the map above show the areas where blizzards happen most often.

Blizzards occur in cold parts of the world. They can happen in northern Asia, Europe, and North America. These areas have cold weather during the winter, and big storms can create blizzard conditions. Blizzards also happen on high mountains on almost every continent. That's because the air is very cold up that high.

A dog copes with the cold by curling up to keep itself warm.

A blizzard's strong winds are dangerous too. They can blow down power lines. Strong winds and low temperatures can cause people and animals to freeze to death. It is best to stay inside and be safe and warm during a blizzard.

This 2011 satellite image shows a major winter storm moving across the midwestern United States.

CHAPTER 2

BLIZZARD SCIENCE

What makes blizzards so scary? The combination of snow, wind, and cold air creates very dangerous conditions.

Long-Lasting Storms

A blizzard lasts for hours. Some blizzards last for days. Blizzards are areas of **low pressure**. That means the air in the atmosphere is rising. High pressure means the air is sinking. The atmosphere has many low- and high-pressure spots. Low-pressure areas bring cloudy weather. High-pressure areas bring clear, sunny weather.

If an area of low pressure meets an area of high pressure, a storm develops. Clouds in low-pressure places become trapped if they meet an area of very high pressure. Very high pressure can stop a storm from moving away. When this happens, a blizzard sits over an area for a long time.

So Much Snow

A lot of snow can fall during a blizzard. That's because the snow comes down so hard. Up to 3 inches (7.6 cm) of snow can fall in just one hour. The snow falls hard because there is a lot of moisture in the clouds.

In February 1978, a blizzard struck the New England area of the United States. The storm dumped between 30 and 40 inches (76 and 102 cm) of snow there. This blizzard is known as the Northeast Blizzard of 1978. It formed when cold air coming down from Canada got stuck over New England. Other weather systems trapped the storm in place. Because the storm could

After the Northeast Blizzard of 1978, people skied on the snow-covered streets of Boston, Massachusetts.

not move away, heavy snow and high winds lasted for a long time. The storm stayed over New England for thirty-six hours.

Roaring Winds

Powerful winds are another ingredient for a blizzard. Wind speeds have to be more than 35 miles per hour (56 kmh) for a storm to be called a blizzard. However, the wind can blow much harder than that.

As winds blow snow around, they can move the snow into huge piles. The piles are called **snowdrifts**. These snowdrifts can be high enough to bury cars. Some

DID YOU KNOW?

A ground blizzard happens when high winds blow snow that is already on the ground.

Skiers struggle to move through blowing snow during a ground blizzard on a mountain in Turkey.

can bury houses! In the Northeast Blizzard of 1978, snowdrifts rose to more than 20 feet (6 meters).

Blinding Conditions

The combination of high winds and heavy snow make it very hard to see. This can create many dangers. Drivers cannot see where they are going. They may hit other cars or drive off the road.

People walking in a blizzard can get lost because they can't see where they are. If people are lost in a blizzard, their body temperature can become low. This is called **hypothermia**. They may also suffer from **frostbite**. Frostbite means parts of the body freeze. A person can freeze to death if they don't find a safe place to go inside.

Poor visibility is dangerous on the water as well. In November 1913, a blizzard hit the Great Lakes. The storm was so fierce it was called the White Hurricane. Ship captains could not see. The storm created winds of 80 miles per hour (129 kmh) and

Slippery road conditions cause many car accidents during blizzards.

Blizzards are dangerous at sea too. They bring strong winds and huge waves, and make it very hard to see.

35-foot (11 m) waves. Twelve ships sank. More than 250 sailors died.

A blizzard that hit the East Coast of the United States in January 2018 created a wind gust of 122 miles per hour (196 kmh) on Mount Washington in New Hampshire. Many other places saw winds of more than 70 miles per hour (113 kmh). Winds from that storm also pushed ocean water into waves of over 50 feet (15 meters).

Some blizzards are so big that they affect huge parts of a country. In March 1993, the Storm of the Century struck the eastern part of the United States. Bad weather stretched from Maine to Georgia. The storm formed when snow and rain moving from the West Coast combined with thunderstorms from the Gulf of Mexico. The storm then hit lots of cold air. This made a lot of snow. The storm was so big that every airport on the East Coast closed. This was the first time that ever happened!

DID YOU KNOW?

In October 2013, a surprise blizzard in South Dakota killed up to one hundred thousand cattle.

HOW DO YOU RATE THAT?

Until the twenty-first century, there was no rating scale for blizzards. In 2004, the National Weather Service created one. The scale is called the Northeast Snowfall Impact Scale, or NESIS. NESIS uses a 1 (Notable) to 5 (Extreme) point scale. It does not rate the storm as it happens. Instead, NESIS measures the effects of the blizzard after the storm has ended. It uses a mathematical formula that includes data on the storm's size, the amount of snow, and the number of people affected by the storm.

Meteorologists use satellites, radar, and computer data to track and predict how a blizzard will behave.

CHAPTER 3

PREPARING FOR AND PREDICTING A BLIZZARD

In the past, it was hard for scientists to **forecast** blizzards. Things have changed a lot over the years. Today, weather predictions help us prepare for big storms.

Staying Safe

Blizzards are dangerous storms. The best way to stay safe is to stay inside. When a blizzard is predicted, most schools close for the day. Many stores and other businesses also close.

Governments may declare a state of emergency during a blizzard. During a state of emergency, many roads are closed. It is against the law for people to drive in a blizzard state of emergency.

Danger can also come from fallen power lines. Lines can fall down when the wind is strong or when too much snow sits on top of the lines. Sometimes trees fall down and pull down power lines. When power lines fall, electricity can go out. The electricity can be out for a few hours or many days after a bad blizzard.

How to Prepare

To prepare for a blizzard, make sure you have lots of food and water at home. You should have enough food and water to last for several days. Remember to have supplies for your pets too.

High winds and heavy snow often bring down power lines.

Your power may go out during a blizzard. Be sure to have batteries, candles, and flashlights in your house. You can use them if the power goes out. Charge all cell phones before the storm hits. Listen to the radio or check the internet for updates about the blizzard.

Technology Helps

Technology is helping make blizzards less severe. Some areas have heated sidewalks that keep snow from sticking. Other areas are experimenting with solar roads. These roads store energy from the sun. They use the stored energy to warm the road during a storm.

Technology can even help us shovel snow! Japanese companies have invented robots that shovel snow. Other

DID YOU KNOW?

Highway departments apply salt to roads before a storm. The salt lowers the freezing temperature of water and helps keep snow from sticking as quickly.

FORECASTS SAVE LIVES

Blizzard warnings help people get ready before a storm hits.

In the past, blizzards often caught people by surprise. The results could be deadly. In 1888, a storm called the Schoolchildren's Blizzard surprised the midwestern part of the United States. At least 235 people were killed. Many were children on their way home from school. Today's weather forecasters use radar and other technologies to predict and track storms. They can warn that a blizzard is coming several days before it happens. These warnings save lives.

Robots that shovel snow could be a great help to people after a blizzard.

similar helpful devices are being thought up and developed today too.

Blizzards of the Future

Scientists believe **climate change** may create more blizzards in the future. Temperatures around the world are rising. Rising temperatures change weather patterns. These changes could bring more severe storms, including blizzards.

Only time will tell what kinds of blizzards we might see in the future. Technology can help us deal with severe storms. So can being prepared.

GLOSSARY

climate change A shift in weather events due to pollution and other factors.

forecast To predict future events.

frostbite Injury to the body from extreme cold.

hypothermia Dangerously low body temperature.

lift When warm air rises over cold air.

low pressure An area in the atmosphere where air rises.

moisture Water in the air.

snowdrifts Tall piles of snow.

visibility The ability to see a certain distance in front of you.

FIND OUT MORE

Books

Meister, Cari. *Disaster Zone: Blizzards*. Minneapolis, MN: Jump!, 2017.

Raum, Elizabeth. *Blizzard!* Mankato, MN: Amicus High Interest, 2017.

Website

Blizzard Facts for Kids

https://kids.kiddle.co/Blizzard

Here you can learn about what causes a blizzard and find information about and photos of some of the worst blizzards ever recorded.

Video

Weather 101: What Is a Blizzard?

https://www.youtube.com/watch?v=uQXtCyoN-tl

Head out into the snow and find out what makes blizzards such amazing storms in this short video.

INDEX

ABOUT THE AUTHOR

Joanne Mattern is the author of hundreds of nonfiction books for children. Storms are one of her favorite subjects to write about, along with sports, history, animals, and biography. Mattern lives in New York State with her husband, children, and an assortment of pets. She has experienced a few blizzards in her life.